AF855042

J 551.5 SIL
Silverman, Buffy.
Investigating weather :
weather systems /

PALM BEACH COUNTY
LIBRARY SYSTEM
3650 Summit Boulevard
West Palm Beach, FL 33406-4198

DO IT YOURSELF

Investigating Weather

Weather Systems

Buffy Silverman

Heinemann Library
Chicago, Illinois

© 2009 Heinemann Library
an imprint of Capstone Global Library, LLC
Chicago, Illinois

Customer Service 888-454-2279
Visit our website at www.heinemannraintree.com

All rights reserved. No part of this publication may be reproduced or transmitted in any form or by any means, electronic or mechanical, including photocopying, recording, taping, or any information storage and retrieval system, without permission in writing from the publisher.

Editorial: Louise Galpine and Rachel Howells
Design: Richard Parker and Tinstar Design Ltd
Original illustrations © Capstone Global Library Ltd
Illustrations: Oxford Designers and Illustrators, Darren Lingard (p. 15), and Geoff Ward (p. 27)
Picture research: Hannah Taylor and Fiona Orbell
Production: Alison Parsons

Originated by Dot Gradations Ltd.
Printed in China by Leo Paper Products Ltd.

13 12 11 10 09
10 9 8 7 6 5 4 3 2 1

Library of Congress Cataloging-in-Publication Data
Silverman, Buffy.
Investigating weather : weather systems / Buffy Silverman. -- 1st ed.
p. cm. -- (Do it yourself)
Includes bibliographical references and index.
ISBN 978-1-4329-2309-9 (hc) -- ISBN 978-1-4329-2316-7 (pb)
1. Weather. 2. Meteorology. I. Title.
QC981.S546 2008
551.5--dc22
2008034935

Acknowledgments
The author and publishers are grateful to the following for permission to reproduce copyright material: © Alamy pp. **4** (Gary Crabbe), **23** (Sami Moudavaris), **38 right** (Utah Images/NASA); © Corbis pp. **6** (Frans Lanting), **18** (TH-Foto/zefa), **27** (Tom Stewart), **31** (Paul A. Souders); © Getty p. **43** (Roger Weber); © Library of Congress p. **38 left**; © Panos Pictures pp. **5** (Morris Carpenter), **26** (G.M.B. Akash), **32-33** (Jocelyn Carlin); © PA Photos p. **39** (PA Archive); © Photolibrary pp. **7** (AlaskaStock), **11** (Christopher Gallagher), **13** (Liysa), **25** (Sandra Middleton), **29** (Franklin Viola); © Science Photo Library pp. **41** (David Hay Jones), **42** (NASA).

Cover photograph of bolts of cloud-to-ground lightning in a supercell thunderstorm over Kansas, USA, reproduced with permission of © Jon Davies (Jim Reed Photography/ Science Photo Library).

The publishers would like to thank Harold Pratt for his invaluable assistance in the preparation of this book.

Every effort has been made to contact copyright holders of any material reproduced in this book. Any omissions will be rectified in subsequent printings if notice is given to the publishers.

Disclaimer
All the Internet addresses (URLs) given in this book were valid at the time of going to press. However, due to the dynamic nature of the Internet, some addresses may have changed, or sites may have changed or ceased to exist since publication. While the author and publishers regret any inconvenience this may cause readers, no responsibility for any such changes can be accepted by either the author or the publishers. It is recommended that adults supervise children on the Internet.

Contents

Any words appearing in the text in bold, **like this**, are explained in the glossary.

What Is Weather?

You and your friends have planned a picnic and are hoping that the day will be sunny and warm. But in the morning, you look outside to see that clouds have gathered and the wind is blowing. Soon the rain starts to fall. The phone rings, and your friend informs you that the picnic has been canceled. Why have your plans changed? The weather! No one wants to sit in the rain and have a soggy picnic. Weather is important when we make outdoor plans.

For many people, weather plays a crucial part in their lives. Farmers need rain so that the seeds they plant will sprout. They need sunshine for their plants to grow. Frosty nights or long droughts can cause them to lose their crops. Pilots and sailors must pay careful attention to the weather to stay safe when they travel. Foresters need to watch the weather when fighting dangerous fires.

When rain pours down, people must change their plans. It's not a good day for a picnic!

The power of weather

Severe weather can cause big problems for people and property. After a huge rainstorm, rivers can overflow their banks and flood nearby neighborhoods. Winds from tornadoes can blow down homes, schools, and other buildings, putting people and other animals in danger. Weather can mean the difference between families having plenty of food and mass starvation. We all depend on the weather.

What causes weather?

Three things cause our weather: heat from the sun, moving air, and water. Our weather is powered by **energy** from the sun. The sun heats the Earth's **atmosphere** and warms its oceans, rivers, lakes, and land. Energy from the sun fuels winds. Winds blow clouds, moving moisture around the globe. In addition, winds blow warm and cool air from one place to another.

Weather can be dry or wet, warm or cold, cloudy or clear, or windy or still. Weather describes the air at a particular place, during a particular time.

When rain does not fall, plants die and farmers lose their crops.

About the experiments

Carrying out the experiments in this book will help you to understand weather. The experiments use simple, everyday materials and tools. Always read through the instructions before you start, and take your time. You will need an adult to help with some of the experiments.

The Power of the Sun

The sun's **energy** powers the weather on Earth. The sun appears small, because it is so far away. But the sun is actually huge. About 109 Earths could fit across the middle of the sun.

The sun is intensely hot. The surface of the sun (the **photosphere**) gives off energy in all directions. Energy from the sun travels to the Earth as light and heat waves. The light we see is part of these waves. Some of the waves are not visible to us. We feel **infrared waves** as heat.

Energy from the sun reaches the Earth's **atmosphere**. The atmosphere is the layer of **gases** surrounding the Earth that we call air. Some energy is absorbed by gases in the atmosphere. Gases and clouds reflect some energy back into space. About half of the sun's energy reaches the Earth's surface, warming the land and oceans. The ground and water give off heat, warming the air above. As the air warms, it begins to move, forming winds that carry weather. The heat is carried upward, and much of it is lost to space.

All year, temperatures are warm in this tropical rain forest.

Even in summer, you can find snow near the North Pole.

On the Earth

The heating of the Earth is uneven. The amount of sunlight that a place receives changes during the day and with the **seasons**. Each day the sun's energy warms the land and oceans. At night the land continues to give off heat.

Some places on Earth receive more of the sun's heat than others. At the **equator**, the sun's rays shine directly over the land and oceans. The heat from the sun is more concentrated along the equator, so temperatures are higher.

At the Earth's two poles, sunlight travels farther through the atmosphere. It is not directly overhead. The sun's rays spread out over a wider area, so they have less heat when they reach land. Temperatures are lower here. Sheets of snow and ice cover the land and parts of the oceans at the poles.

In the atmosphere

Weather takes place in the Earth's atmosphere. Gases in the Earth's atmosphere stay close to Earth because of the pull of Earth's gravity.

Two gases make up most of our atmosphere. Air is about 78 percent nitrogen and 21 percent oxygen. The remaining one percent is made up of small amounts of **water vapor**, hydrogen, carbon dioxide, and other gases.

Gases in the atmosphere protect life on Earth by keeping out the sun's harmful rays. They also trap some of the heat from the sun during the day, keeping the Earth warm at night. Without the atmosphere, there could be no life on Earth.

Layers of the atmosphere

Scientists divide the atmosphere into five layers. The layer closest to the Earth is called the **troposphere**. The troposphere has water vapor and clouds that produce our weather. It is about 7 kilometers (4.3 miles) thick at the poles and 17 kilometers (10.5 miles) thick near the equator. Air in the troposphere is warmest near the Earth's surface. The sun warms the oceans and land. Heat rises from the Earth's surface and warms the air above it. The temperature of air decreases farther up in the troposphere. Water vapor and carbon dioxide in the troposphere trap the sun's heat, keeping the Earth warm.

Going higher ...

The air above the troposphere is thinner and does not have enough oxygen to breathe. This next layer is called the **stratosphere**. The stratosphere contains a layer of gases called the **ozone** layer. Ozone absorbs harmful ultraviolet rays from the sun that would otherwise burn our skin. Jets often fly in the stratosphere because its air is stable.

...And higher

Above the stratosphere is the **mesosphere**, the coldest layer of the atmosphere. The layer above the mesosphere is called the **thermosphere**. Its temperatures can reach up to 1,500 °C (2,730 °F). Space shuttles orbit here. Past the thermosphere is the **exosphere**. The air in the exosphere is very thin and gives way to space.

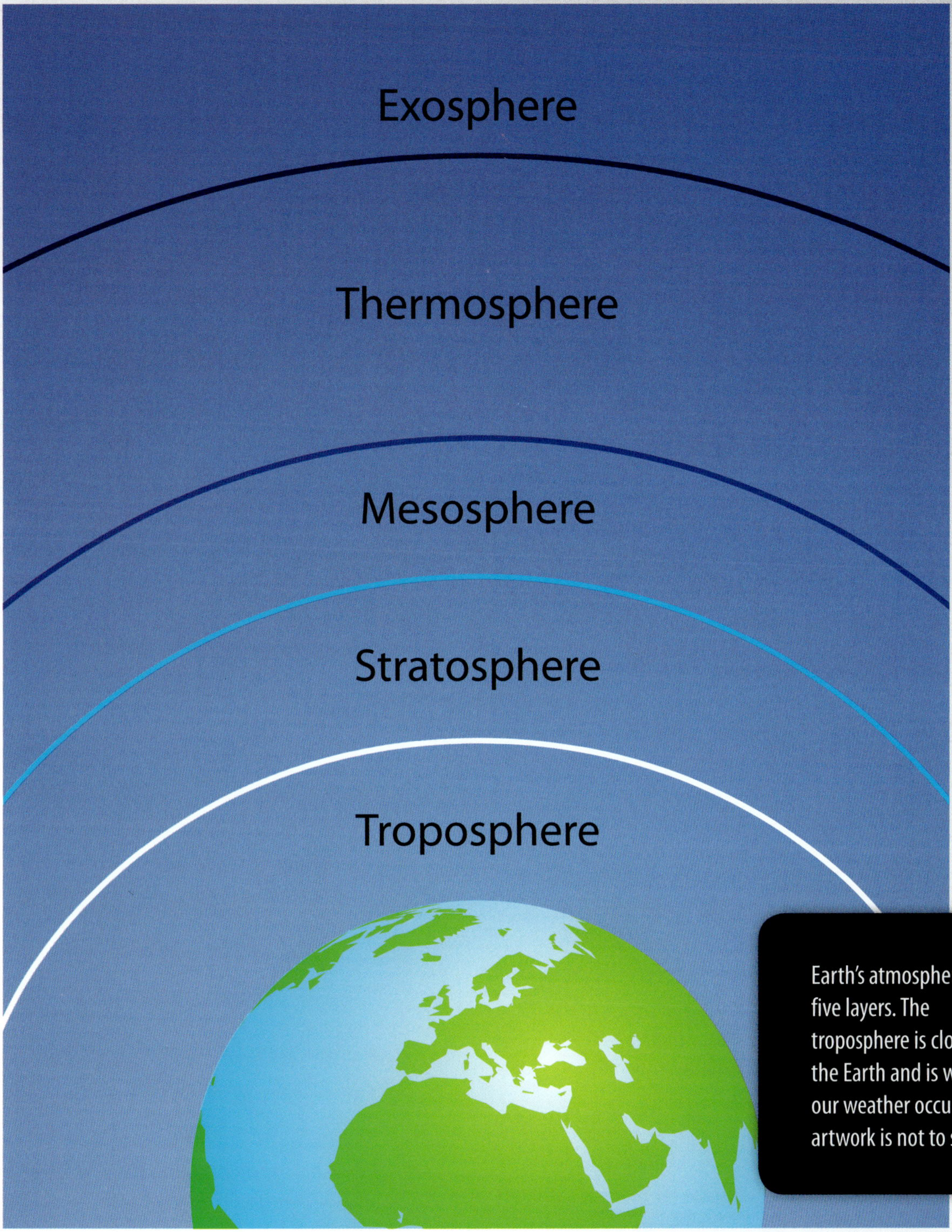

Earth's atmosphere has five layers. The troposphere is closest to the Earth and is where our weather occurs. (This artwork is not to scale.)

Sunlight around the globe

For this activity you will need:

* A globe or balloon
* A flashlight.

Steps to follow

1 A globe can show you how sunlight reaches the Earth. If you do not have a globe, blow up a balloon. Draw a line around the middle to represent the equator.

2 Darken the room. Put the globe on a table, or ask a friend to hold the balloon steady, with the equator around the middle.

3 Use a flashlight to represent the sun. Shine the flashlight directly on the equator. Notice the size and intensity (strength) of the light beam.

4 Keeping the flashlight horizontal and the same distance from the globe, move it up so it shines on the North Pole. Does the flashlight light a larger or smaller area? Is the light more or less bright?

Notice how the flashlight lit a smaller area when shining directly on the equator. The spot of light was brighter. When the flashlight shone on the polar area, the light spread over a larger area. The light beam was dimmer.

Climates around the globe

We saw in the experiment that because the Earth's surface is curved, different places on Earth receive different amounts of heat energy. The sun shines directly on the equator. Intense sunlight makes areas near the equator very warm. The farther a place is from the equator, the less heat it receives from the sun. The coldest regions on Earth are farthest from the equator. Although weather changes from day to day, an area has similar patterns of weather each year. This pattern of weather is called **climate**. Near the equator, many places have hot, tropical climates. Polar regions have cold, polar climates.

The climate of a region also depends on the amount of rainfall it receives. Places with desert climates receive little rainfall because of the dry air.

Places with temperate climates are located between the polar regions and tropical areas. Temperate climates are warm in summer and cold in winter.

Land and sea

For this activity you will need:

* Two identical plastic containers (same size and color)
* Soil or sand
* Water
* Two thermometers
* A marker
* A ruler
* A notebook to record results.

Steps to follow

1. Measure to 2.5 cm (1 inch) below the top of each container. Mark a line at this height.
2. Fill one container with water to the marked line.
3. Fill the other container with soil or sand to the marked line.
4. Place a thermometer in each container, and record the temperature.
5. Place both containers in a sunny spot for one hour. Record the temperature.
6. Place both containers in the shade for one hour. Record the temperature.

Which container warmed fastest in the sun? Which container lost its heat more quickly in the shade? You probably found that soil warms up and cools down more quickly than water. While water takes longer to heat, it holds on to that heat for a longer time.

On a summer's day, an ocean breeze keeps the beach cool.

Oceans and land

Like the water in the experiment, oceans take longer to warm and cool than land. Because they do not change temperature as quickly as land, oceans **moderate** the climate. A warm ocean keeps the air warm in the winter. In the summer, the ocean keeps the air cooler. On a cold winter's day, the temperature along the coast may be warmer than farther inland. In the summer, you can go to the beach to enjoy cooler weather. The climate of coastal lands is often cooler in summer and warmer in winter.

Ocean **currents** can also change the climate of land near the coast. Ocean currents are flowing streams of cold or warm water. Winds blow ocean waters. Because of the way the Earth rotates, these currents move in circular paths. Norway's coast extends close to the North Pole. But its climate is milder than you might expect because of the warm ocean currents that flow past.

Our Spinning Earth

A blast of wind whips snow through the air. You pull up your hood and hurry inside, eager to get out of the cold. Winter seems like it will never end. Six months later, the hot sun beats down on you. You wipe away sweat as you race down the soccer field. On a long summer's day, you might wish that summer would never end.

The weather changes during different periods of the year. We call these changes **seasons**. You might expect hot weather on a summer afternoon, and a thunderstorm on a summer night. In the winter you might look forward to a snowstorm and a day of skiing or sledding. Each season brings different weather. In some tropical areas there are just two seasons: rainy and dry.

The Earth tilts

Every year the Earth revolves one time around the sun. As the Earth moves, it is tilted at an angle. Because the Earth is tilted on its **axis**, we have seasons.

If you live in the northern **hemisphere**, then you enjoy long summer days in July. At this time of year, the northern half of the Earth is tilted more toward the sun. The southern hemisphere is tilted away from the sun. It receives less sunlight, and has cold winter days.

Six months later, the Earth has completed half its orbit around the sun. Now the northern hemisphere is farther from the sun. The days are short, and it is winter. The southern hemisphere is tilted more toward the sun. There the days are long and warm, and it is summer.

The seasons change according to Earth's position in its orbit around the sun.

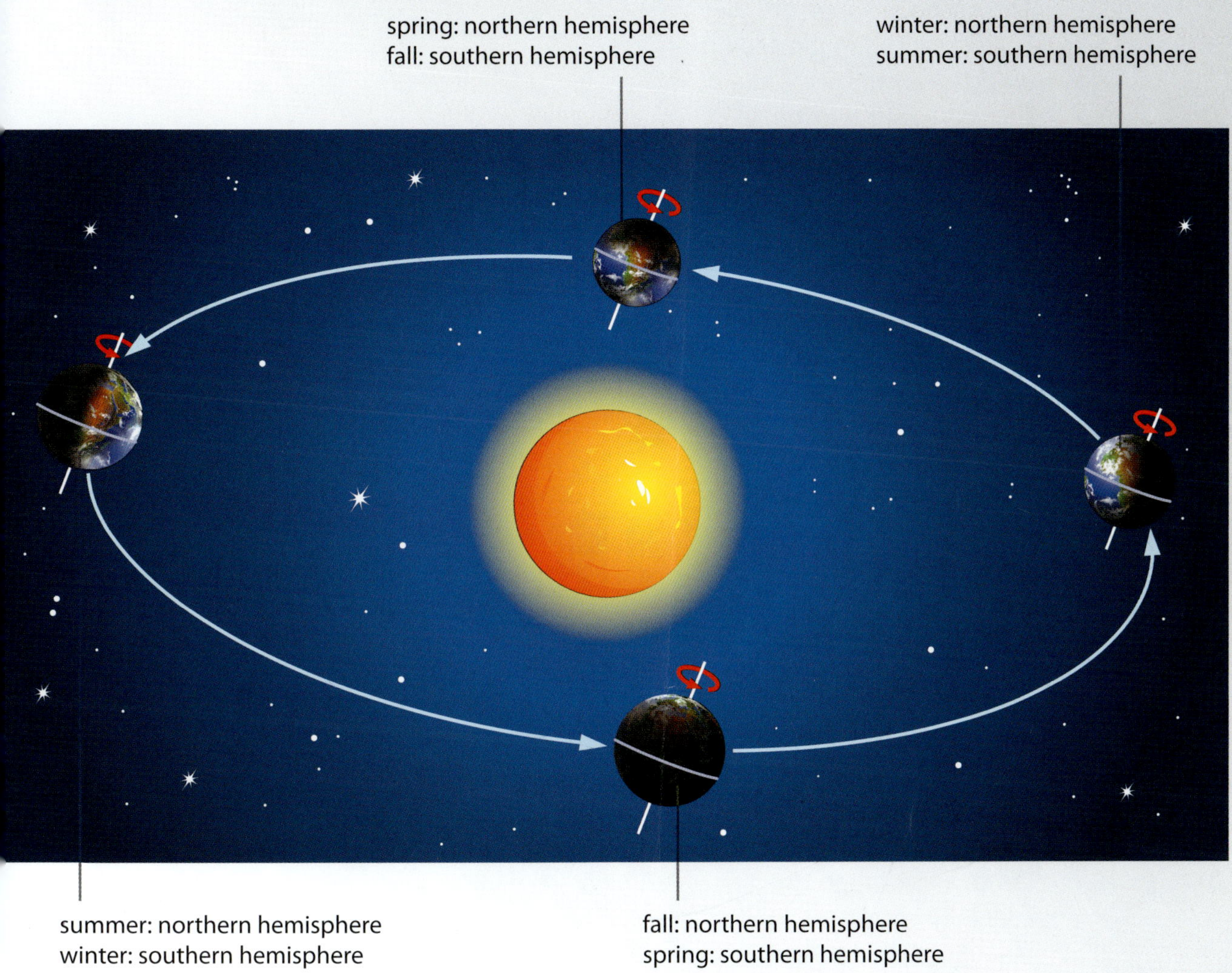

Day and night

Every 24 hours the Earth turns around once on its axis. When the place where you live faces the sun, it is daytime. As the Earth continues to spin, the sun sets and night begins.

Reasons for seasons

For this activity you will need:

* An orange
* A metal skewer
* A marker
* A lamp.

Warning: Adult help will be needed for this experiment.

Steps to follow

1 The orange represents the Earth. Use the marker to draw a ring around the middle of the orange (the **equator**). Mark an "N" on top (the North Pole) and an "S" on the bottom (the South Pole). Draw an "X" to show about where you live on Earth (north or south of the equator).

2 Ask an adult to put a skewer through the orange, piercing the "N" and "S" marks. The skewer represents the Earth's axis.

3 With the help of an adult, put the lamp on a table in the middle of the room. Remove the lampshade, turn on the lamp, and darken the room. The lamp represents the sun.

4 Hold the orange by the skewer, with "N" on the top and "S" on the bottom. Tilt the skewer (axis) slightly so that "N" is pointing towards the lamp. This represents summer in the northern hemisphere and winter in the southern hemisphere. Notice how much of the Earth is lit above the equator, and how much below the equator. Which half of the Earth receives more light?

5 Without changing the tilt of the Earth, slowly spin it so the "X" is facing the lamp. It is now day where you live. Continue to spin the orange slowly until the "X" faces away from the lamp. It is now night where you live.

6 Keeping the tilt of the orange the same, bring the orange to the opposite side of the lamp. The "N" is now tilted away from the lamp, and the "S" is tilted towards the lamp. It is winter in the northern hemisphere and summer in the southern hemisphere. Notice the portion of Earth that is lit up above the equator and below. Spin the orange to show day and night.

7 Bring the orange halfway between the summer position and winter position. Which season does this represent? How does the light compare at the two poles?

On the Move

Look up at the sky. Does the sun shine brightly? Do clouds race overhead? Moving air can carry warm, moist air. It can bring cold, dry air. As air flows from one place to another, weather changes. A sunny day can quickly turn into a stormy one.

We call moving air "wind". Although you cannot see the air move, you can see the impact of wind. A gentle wind stirs leaves and branches, while powerful winds can knock over trees and lift roofs off.

Kites perform better on a windy day, flying high in the sky.

What causes wind?

Wind is caused by the uneven heating of the Earth's surface. Oceans, lakes, and rivers cover more than two-thirds of the Earth's surface. The rest is covered by land, including mountains, valleys, and plains. Bodies of water and different landforms absorb the sun's **energy** at different rates. During the day, air over the land heats up faster than air over the water. Warm air is lighter than cold air, and rises. As warm air rises, cool air rushes in to take its place. This movement creates wind.

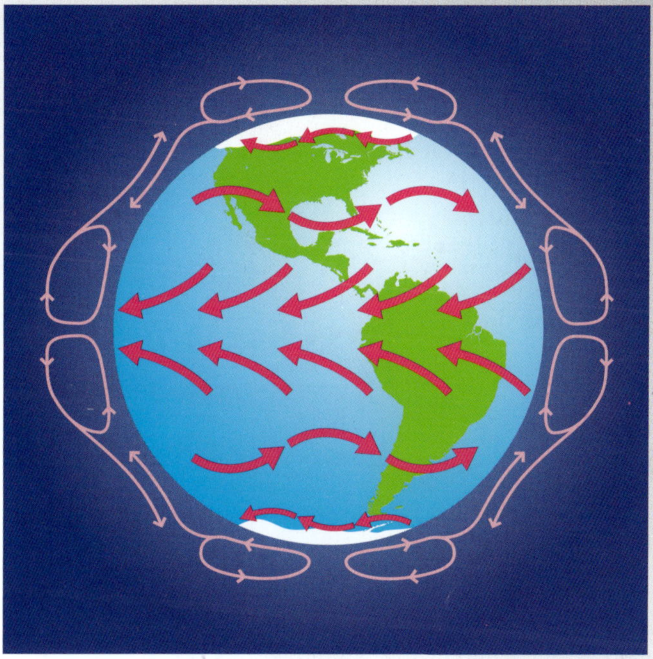

Winds circle around the Earth.

Wind patterns

Wind is described by its speed and direction. The speed and direction of wind can change from day to day. But there are large wind patterns around the globe that help people predict the weather.

The land near the **equator** rises to a higher temperature than land near the poles. This heating difference sets up global wind patterns. As air is heated, it moves north or south of the equator. The warm air starts to cool and sink. Some cooling air moves back to the equator. At the North and South Pole, cold air sinks. As air moves away from the poles, it warms and rises, and some moves back to the poles. Air between the equator and the poles is pushed by moving air on both sides.

Blowing in the wind

For this activity you will need:

* A lamp with shade removed
* Talcum powder
* A towel.

Steps to follow

1. Ask an adult to help you remove the lampshade from a lamp.
2. Place a towel under the lamp to catch spills.
3. Turn on the lamp. Wait for two minutes for the bulb to heat up.
4. Sprinkle a small amount of powder on the bulb from above.
5. Observe what happens to the powder.

Warning: Adult help will be needed for this experiment.

Making wind

For this activity you will need:

* A lamp with shade removed (as on page 20)
* A piece of scrap paper, about 10 cm (4 in.) square
* A pencil
* Scissors.

Steps to follow

1 Draw a spiral on the scrap paper.

2 Cut along the spiral line with the scissors.

3 Turn on the lamp, and wait for two minutes for the bulb to heat up.

4 Balance the center of the cut paper carefully on the pencil. Do not make a hole in the paper. The paper will spiral down around the pencil.

5 Hold the pencil just above the bulb.

6 Observe what happens to the spiral.

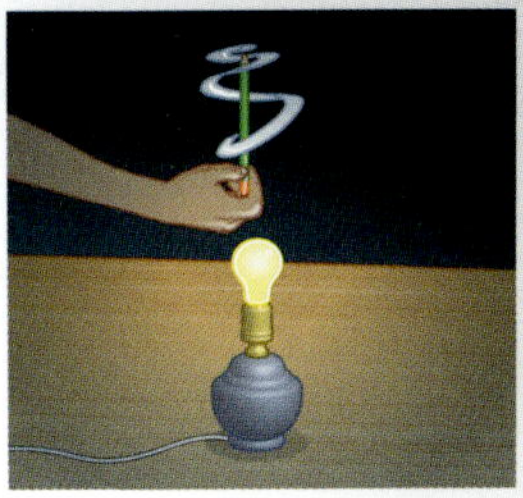

When you turned on the lamp, the bulb warmed the air around it. In the same way, the sun warms the surface of the Earth, and the land or water warms the air above it. As air heats, it rises. Although you cannot see moving air, you can see anything that it carries. By sprinkling powder on the bulb, you can observe the movement of air. A cloud of powder rises with the warm air. Cool air is drawn in to replace the warm air, and wind is created.

You also observed air movement when you balanced the spiral paper above the lamp. Again, the bulb heated the air around it, causing the warm air to rise. Cool air was pulled in to replace the warm air. The moving air pushed against the paper, and the spiral began to spin.

Make a weather vane

For this activity you will need:

* Scissors
* Cardboard
* A plastic water bottle
* A straw
* Gravel or sand
* A marker
* A compass
* Paper and a pencil.

Steps to follow

1 With your marker, draw an arrow with a long tab on a piece of cardboard.

2 Using the scissors, cut out the arrow with the tab.

3 Carefully insert the tab into the straw. The tab sticks into the straw and the arrow rests on top of the straw.

4 Draw a line about 4 cm (1.6 in.) above the bottom of the plastic bottle. Fill the bottle to the line with gravel or sand. The weight of the gravel will keep the bottle from blowing over in the wind.

5 Place the straw in the bottle, with the arrow resting on its top. The straw must be above the level of the gravel, so it can move freely. If necessary, trim the straw.

6 Using a marker, mark "N" (for north) on one side of the bottle. Mark an "S" on the opposite side. With "N" facing away from you, mark "E" on the right side of the bottle. Mark a "W" on the left side of the bottle. Your weather vane is complete.

7 Take the weather vane and the compass outside. Set the weather vane on a high, open place (for example, a play structure). Using the compass, locate which direction is north. Place the weather vane so "N" faces north. Watch the arrowhead spin. When it stops turning, it will point into the wind.

8 Record the direction from which the wind is coming. Repeat several times on a windy day. Repeat also on a day with only light wind.

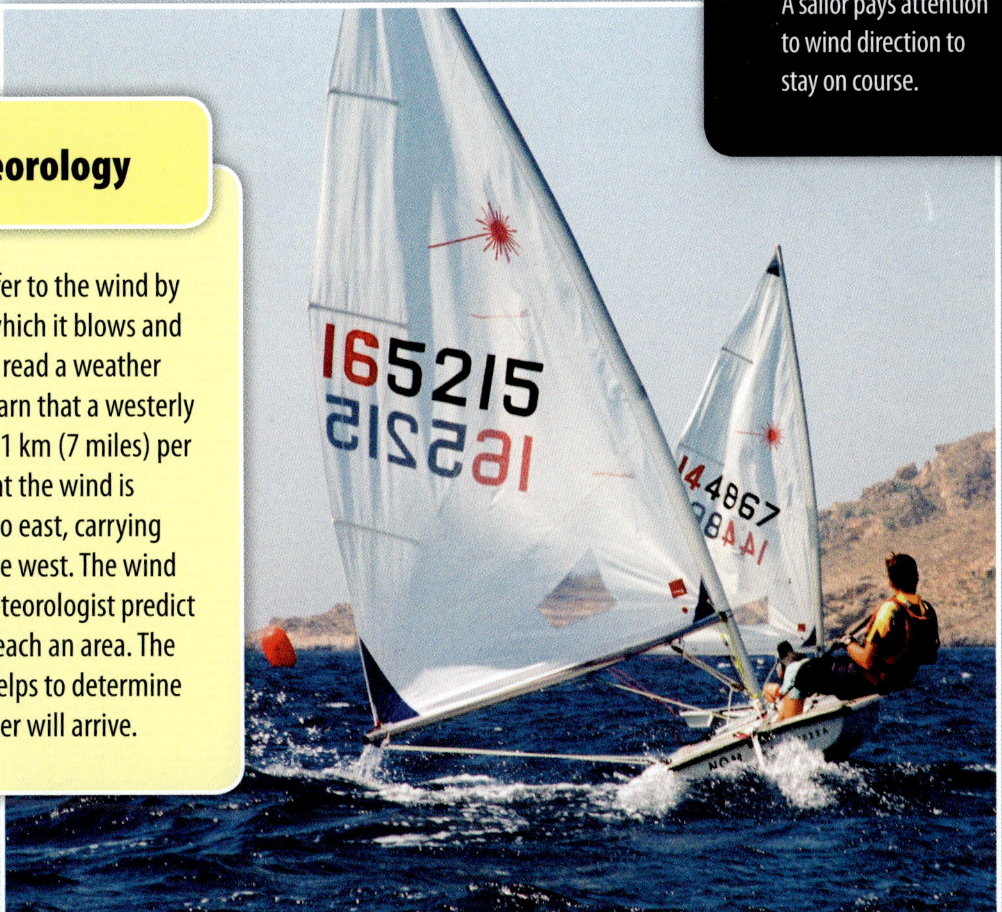

A sailor pays attention to wind direction to stay on course.

Wind and meteorology

Meteorologists refer to the wind by the direction from which it blows and its speed. When you read a weather report, you might learn that a westerly wind is blowing at 11 km (7 miles) per hour. This means that the wind is blowing from west to east, carrying the weather from the west. The wind direction helps a meteorologist predict what weather will reach an area. The speed of the wind helps to determine how fast that weather will arrive.

Steps to follow

Measuring wind speed

For this activity you will need:

* A paper plate
* Four paper cups with handles, all the same size
* A marker
* A ruler
* A thumbtack
* A pencil with an eraser
* Double-sided tape
* A watch with a second hand.

1 An **anemometer** measures wind speed. Make an anemometer to measure wind speed near your home. Mark one cup by coloring the outside with a colored marker.

2 With a pencil and ruler, draw two lines across the plate. The lines meet at the center of the plate.

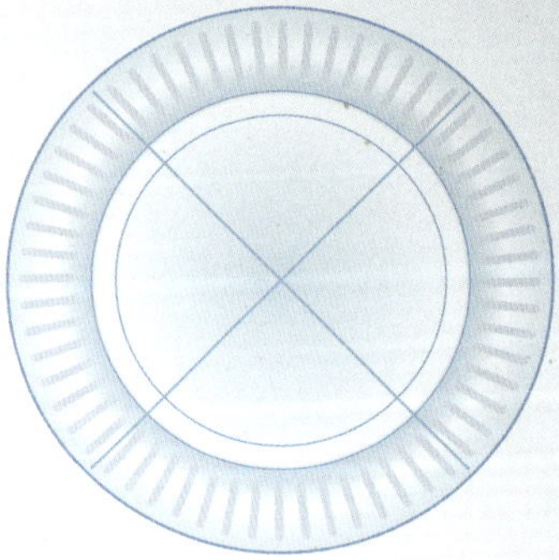

3 Place a piece of tape on the side of each cup. Stick the cups to the edge of the plate, with all cups facing the same direction.

4 Push the thumbtack through the center of the plate. Stick it into the eraser-end of the pencil.

5 Take your anemometer outside on a windy day. When it is spinning, count the number of times that you see the marked cup go around in one minute. That is the wind speed in revolutions per minute. Record the wind speed, and note the date and time.

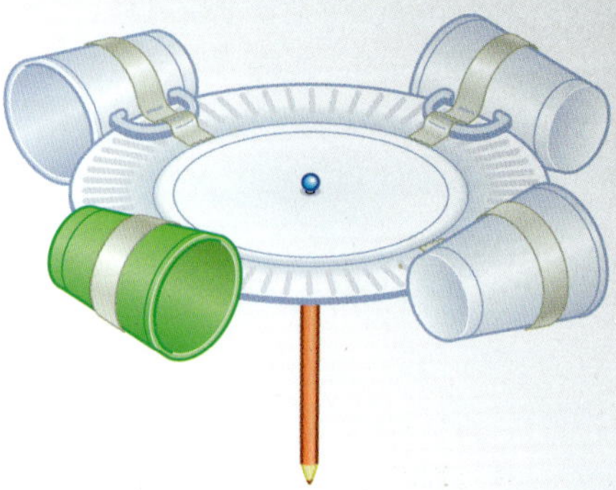

6 Measure the wind speed at different times of the day. Does the wind speed change? Record wind speed for one week. Does the speed change from day to day?

Types of winds

Meteorologists report wind speeds in kilometers or miles per hour. They classify winds according to their speeds. The gentlest winds (1–5 kph; 1–3 mph) are called light air. A wind of 6–11 kph (4–7 mph) is called a light breeze. Leaves rustle during a light breeze. A wind of 39–49 kph (25–31 mph) that bends large branches is called a strong breeze. Gale winds are more powerful. Tree branches break during a severe gale of 75–88 kph (47–54 mph). The most powerful winds are hurricane winds. At more than 118 kph (73 mph), they can destroy buildings and uproot trees.

Record winds

In 1934 the fastest wind recorded with an anemometer was measured on top of Mount Washington, New Hampshire. It blew 372 kph (231 mph). In 1999 a tornado's winds in Oklahoma were measured at 512 kph (318 mph). They were measured with a truck-mounted **doppler radar**.

Hurricane winds can destroy homes.

It's Raining, It's Pouring!

When you think of weather, you might picture rain, snow, sleet, hail, and clouds. Water falling from the sky is called **precipitation**. Where does precipitation come from? It is part of the **water cycle** that carries water around the Earth.

The sun shines on oceans, rivers, lakes, and puddles. The sun's **energy** heats water. When water is heated, some of it **evaporates**. It changes from a **liquid** to an invisible **gas** called **water vapor**.

Rain falls when water droplets in clouds grow.

In the sky

Warm air mixed with water vapor rises in the sky. Warm air can hold more water vapor than cool air. As it rises, cooler temperatures turn water vapor back into liquid water. This change from gas to liquid is called **condensation**. In the sky tiny droplets of water condense on particles of dust. If it is very cold, water vapor may change into ice crystals. The droplets and ice crystals form clouds. The droplets reflect sunlight, making clouds appear white. Droplets are small and light, so they stay suspended in the sky.

Precipitation

The heat of the sun makes some of the droplets in a cloud evaporate. Warm air in clouds rises with water vapor. As it rises, the air cools, more droplets condense, and the cloud grows. As more and more droplets condense, the cloud becomes heavier and heavier. Now the droplets are packed so closely together that light cannot shine through, and the cloud looks dark. When the droplets grow too heavy to stay suspended they fall to the Earth as rain, snow, sleet, or hail.

On a cold winter's day, precipitation falls as snow.

Water spends about 10 days in the air before falling to the ground again.

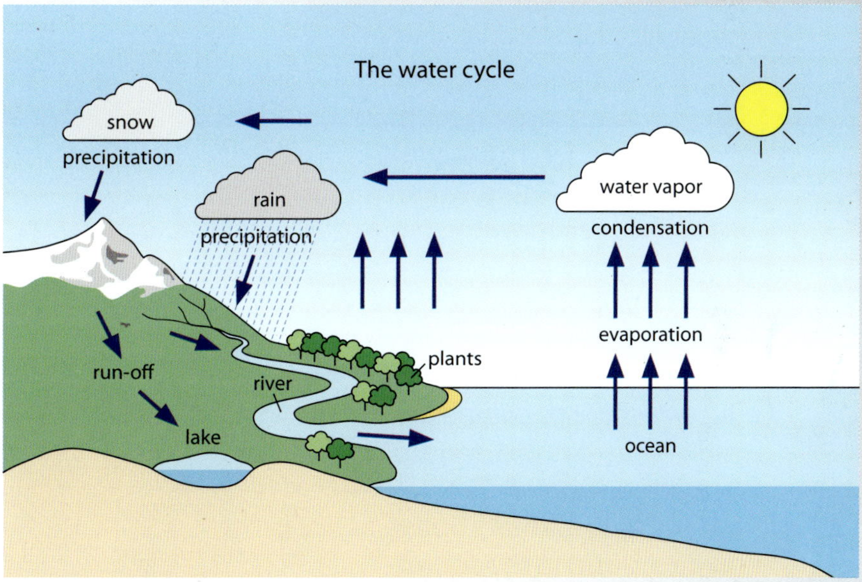

Water continuously travels around the Earth. Some of the water that falls as rain or snow seeps underground. Plants and animals use some. Some water joins rivers and lakes. Eventually, it evaporates and journeys back to the sky to repeat the cycle again. The water that falls as snow or rain today is the same water that was on the Earth when life began.

Making a cloud

For this activity you will need:

* Ice
* A metal pan
* Warm water
* A clear glass jar.

Steps to follow

1 Fill the metal pan with ice. Leave it for five minutes, so the pan becomes very cold.

2 Pour 2.5 cm (1 in.) of warm water into the clear jar.

3 Set the metal pan with ice on top of the jar. Observe what happens inside the jar.

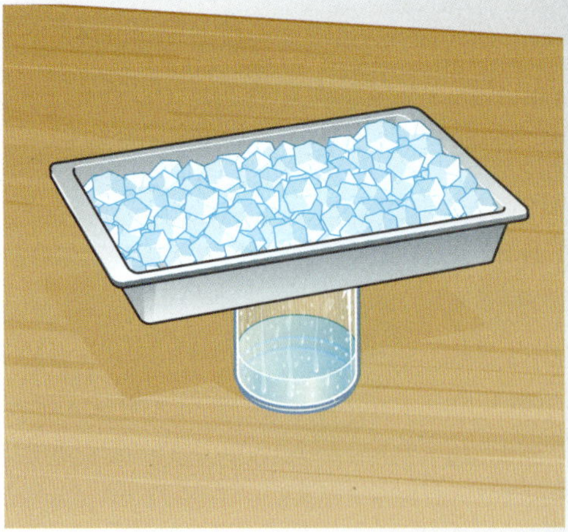

When you filled the jar with warm water, some of the liquid water evaporated into the air in the jar. The water changed from liquid water into a gas, called water vapor. As the warm air rose to the top of the jar, it carried water vapor with it. When the air reached the cold metal pan, the water vapor cooled. As it cooled, the gas condensed, forming tiny droplets of liquid water. These water droplets made a cloud at the top of the jar, similar to ones formed in the sky.

Water vapor

All air contains water vapor. Set a glass of ice water on a table, and you soon notice that the outside of the glass is wet. Water vapor in the air around it has condensed on the cool glass. On a hot, sticky day, you can feel the water in the air. The air is **humid**, because it contains a lot of water vapor. Because warm air is rising and in motion, it can hold more water vapor than cold air.

Air that has as much water vapor as it can hold is called **saturated air**. When it can hold no more water vapor, the vapor in saturated air begins to condense, turning into water droplets. If water vapor condenses near the ground, it forms fog or mist. On a cool morning, you might see fog forming above a lake, as water vapor rising from the lake cools and condenses. When water vapor condenses high above the ground, it forms clouds.

Cloud cover

Clouds come in three main shapes: cirrus, cumulus, and stratus. High in the sky are thin, wispy **cirrus clouds. Cumulus clouds** are fluffy, white clouds that can grow as they rise and bring rain. **Stratus clouds** form flat blankets and are low in the sky.

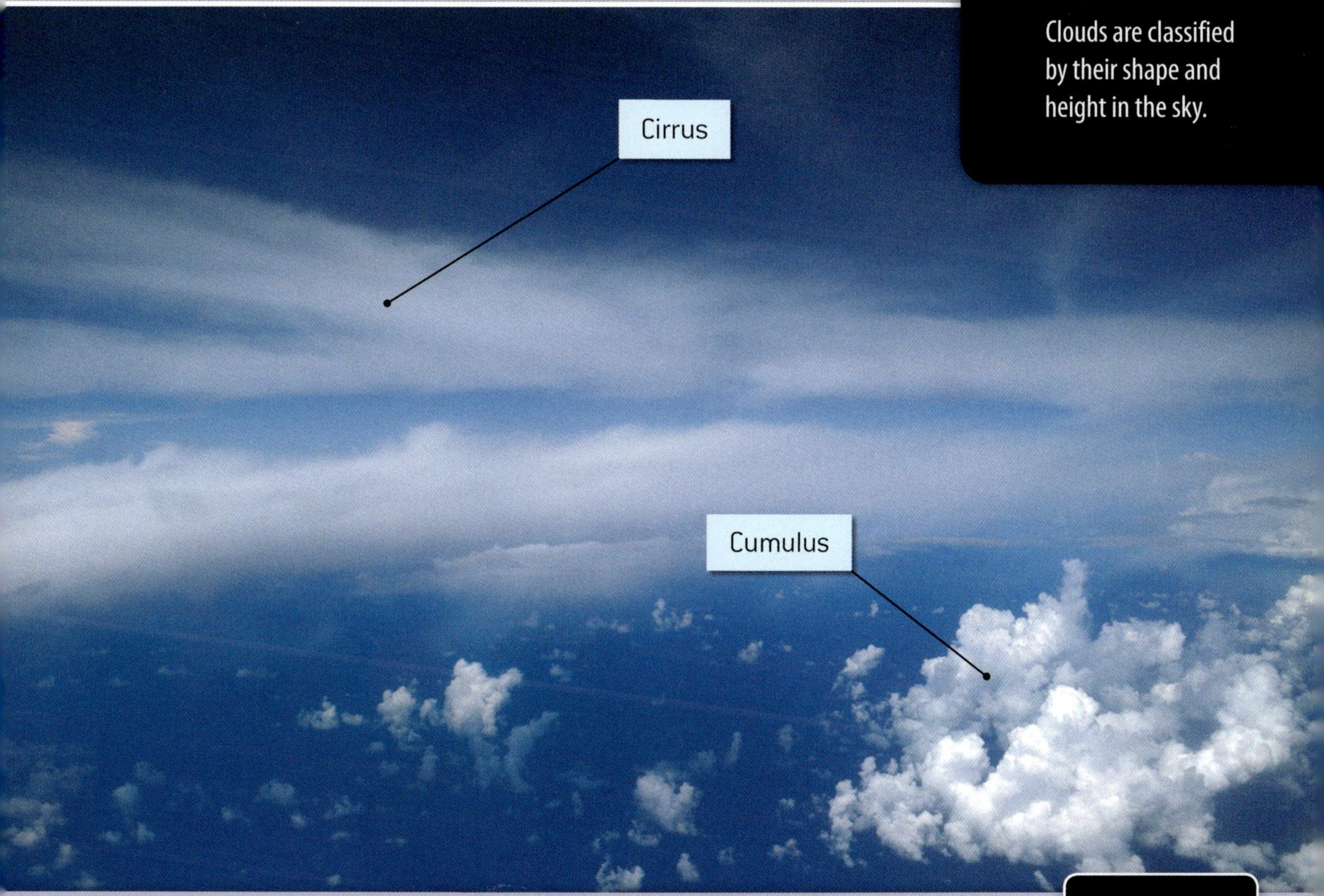

Clouds are classified by their shape and height in the sky.

Wet weather

For this activity you will need:

* A large plastic bottle (2 liter/ 68 fl. oz.)
* Scissors
* Colored tape
* A ruler
* Gravel
* Water
* A notepad and a pencil.

Steps to follow

1 Build a **rain gauge** to measure rainfall. With the help of an adult, cut off the top portion of the plastic bottle. Make your cut just below where the sides slope. Keep the cut portion of the bottle.

2 Fill the base of the bottle with about 2.5 cm (1 in.) of gravel.

3 Place a strip of colored tape on the outside of the bottle, above the gravel.

4 Fill the bottle with water to the top of the tape. You will measure rainfall above this level.

Warning: Adult help will be needed for this experiment.

5 Turn the cut top of the bottle upside down. Push it inside the lower portion of the bottle. The top portion forms a funnel to collect rain.

6 Place the rain gauge in an open area where there are no trees or buildings that can block rainfall.

7 Measure the amount of rainfall at the same time each day. Using a ruler, measure the water level above the tape. That will tell you how much rain has fallen. Record the date and the amount of rainfall. Empty the water as far as the tape line after measuring.

Rain

Inside a cloud, droplets condense on tiny dust particles. Tiny droplets are blown together and join up, forming larger droplets. When they grow too heavy for the cloud to carry, they fall. As they fall, more vapor condenses on them, making larger raindrops.

Snow

When the temperature inside a cloud is very cold, ice crystals form. Water droplets freeze on the crystals. Many crystals stick together, growing into snowflakes that fall to the ground.

Hail

Hail can form during thunderstorms. Ice crystals in towering clouds are blown up and down. As air **currents** push the crystals up and down, water freezes on them, and they grow. When hailstones fall, they can be larger than golf balls.

Sleet

On a cold day, raindrops might freeze as they fall, covering the ground with a blanket of wet, sloppy sleet.

In a rain forest, rain falls almost every day. The wettest places on Earth receive more than 13 meters (42 ft.) of rain per year.

Changing Weather

When you wake up, the sky is clear and the sun shines brightly. But the **meteorologist** on TV says that a storm is heading your way. Soon the sky darkens, thunder rumbles, and lightning flashes. Why is the weather constantly changing?

Air has weight as it presses down on the Earth. This is called **air pressure**. High and low pressure systems are due to the movement of air around the Earth. A **high pressure** area has more air over a location on the Earth, creating more weight, or pressure. **Low pressure** areas are the opposite.

Moving molecules

Gases in air are made of tiny particles that you cannot see, called **molecules**. The molecules in a gas are always moving. When they are heated, the molecules get more **energy** and move farther apart and faster. This makes heated air less **dense**, and it rises. A hot air balloon rises to the sky because the balloon is filled with hot air that is less dense than the surrounding air.

One source of circulating air is the hot air that rises near the **equator**. This hot air rises and moves away from the equator to the north or south. At the other extreme is the cold air near the North and South Poles. This air has more density and slides under the warmer air from the equator. When the cold air meets the warm air they mix in a circular motion. This is caused by differences in temperature and the turning of the Earth.

High and low pressure areas

In low pressure areas the circular motion is counter clockwise, pulling air into the center of the area like a whirlpool. This circulating pattern of air pulls in the **water vapor** from around it. As the air gets to the center it has to rise. Air cools as it rises, and clouds form. Water droplets in clouds collide and join together, becoming heavier until **precipitation** falls. Low pressure leads to cloudy weather, gray skies, and rain.

High pressure areas have a counter clockwise circulation that pushes the air out from it, "clearing" the area of moisture and weather. High pressure areas usually have clear skies and good weather.

When air rises it cools, leaving an area of low pressure. This can lead to a storm.

Expanding air

For this activity you will need:

* A plastic water bottle with a narrow opening
* A freezer
* A balloon
* A large saucepan
* Hot water.

Steps to follow

1 Put the water bottle in the freezer. Wait for 15 minutes so that the air inside the bottle cools.

2 Fill the saucepan with hot water.

3 Take the empty bottle out of the freezer. Carefully stretch the balloon over the opening of the bottle.

4 Place the bottle in the saucepan of hot water, holding it upright. Watch what happens to the balloon.

5 With the balloon still on the bottle, return it to the freezer for another 15 minutes.

6 Remove the bottle from the freezer. What has happened to the balloon?

When you placed the bottle in the freezer, the air inside it cooled. After removing the bottle from the freezer and placing it in hot water, the cold air started to warm. As the air warmed, molecules in it moved in all directions. The air expanded and inflated the balloon.

When you returned the bottle to the freezer, the air inside it cooled again. The molecules in the air in the bottle slowed down and became denser. The air molecules sank, and the balloon deflated.

Over land and sea

As you saw in this experiment, air is warmed when it is placed in hot water. In the same way, air above oceans or land is warmed. The sun heats oceans and land, which then warm the air. This warm air has high air pressure. Its molecules have more energy and move in all directions. Air that is trapped in a balloon makes the balloon expand. But the air outside is not trapped. It rises to the sky, carrying water vapor with it.

Cooling air makes the air more dense. Air sinks toward the land, creating a high pressure area. Because air is drawn from an area of high pressure to a region of low pressure, wind is created.

Make a barometer

For this activity you will need:

* A large jar
* A large balloon
* Two drinking straws
* Sticky tape
* Scissors
* A ruler
* A notepad and a pencil.

Steps to follow

1 A **barometer** measures air pressure. Make a barometer and track changes in the air pressure.

2 With scissors, cut off the neck of the balloon. Throw the balloon neck away.

3 Stretch the remaining part of the balloon over the opening of the jar. Tape it in place, so no air can enter the jar.

4 Cut a small slit in the end of one straw. Squish the cut end and push it inside the other straw, forming one long straw. Tape the spot where the two straws join.

5 Tape one end of the long straw flat to the middle of the balloon. Your barometer is complete.

6 Place the barometer on a shelf where it will not be disturbed.

7

To measure air pressure, hold the ruler upright on the shelf, next to the far end of the straw, with the numbers rising. Record the height that the straw points to.

8

Over the next two weeks, measure and record the air pressure at the same time each day. Note whether the pressure goes up, down, or holds steady. Note the weather conditions at the time of your recording (for example sunny, fair, overcast, rainy, stormy).

How your barometer works

When air temperatures rise, the molecules in air move faster. There is low air pressure, and the air inside the barometer jar expands and rises. The warm, light air pushes the balloon up. Like a seesaw, the far end of the straw drops. The number you record is lower, indicating low air pressure.

When air cools it sinks, creating high air pressure. The sinking air pulls the balloon on the jar downward, moving the straw. The far end of the straw points upward, and you record a high number.

You probably found that when you recorded lower air pressure, the weather was cloudy, and perhaps rain fell. As the air pressure increased, there were fewer clouds. The weather was sunny and warm.

Forecasting the Weather

For thousands of years, people have attempted to predict the weather. Weather forecasts were once based on watching nature. People observed the shape of clouds, the color of the sky, haloes around the moon, and when plants bloomed.

In 1837 the invention of the telegraph allowed forecasting to become a science. Weather stations could now report conditions from a wide area. As the number of weather stations increased, predictions became more accurate.

The invention of the telegraph meant that people could learn of weather conditions in many different places.

Today, **meteorologists** use **satellites** to predict weather. Satellites give pictures of cloud formations over a large area. They are constantly updated. Data about temperature, wind speed, clouds, and **air pressure** are collected. Computers use the data to predict weather. Forecasts are sent over television and the Internet. Warnings are given when severe storms approach.

Satellite pictures help meteorologists predict when a storm will arrive.

Reading weather maps

Meteorologists use weather maps to track weather changes. Look in a newspaper at a weather map, and try to find the following:

Isobars are lines connecting points where the pressure is the same. Rings of isobars show **high-pressure** and **low-pressure** zones (marked with an H or L). These lines have a number on them showing the air pressure. A high pressure zone will usually be cloudless and sunny. A low pressure zone might have rains and wind.

Air travels in large masses of the same temperature and moisture. The place where two **air masses** meet is called a front. Symbols on a weather map show fronts:

Blue triangles. These show a **cold front**, where a cold air mass pushes against a warm one and replaces it. The warm air quickly rises to the sky, forming clouds and wind. A cold front can produce clouds and storms, followed by clear, cooler weather.

Red semi circles. These show a **warm front**, where a warm air mass pushes over a cold air mass. Warm fronts move slowly, bringing cloudy weather and rain.

Red semi circles alternating with blue triangles. An **occluded front** shows where a fast moving cold front overtakes a warm front. The cold air lifts up the warm air mass, pushing it higher in the sky. Heavy rains and thunderstorms may occur.

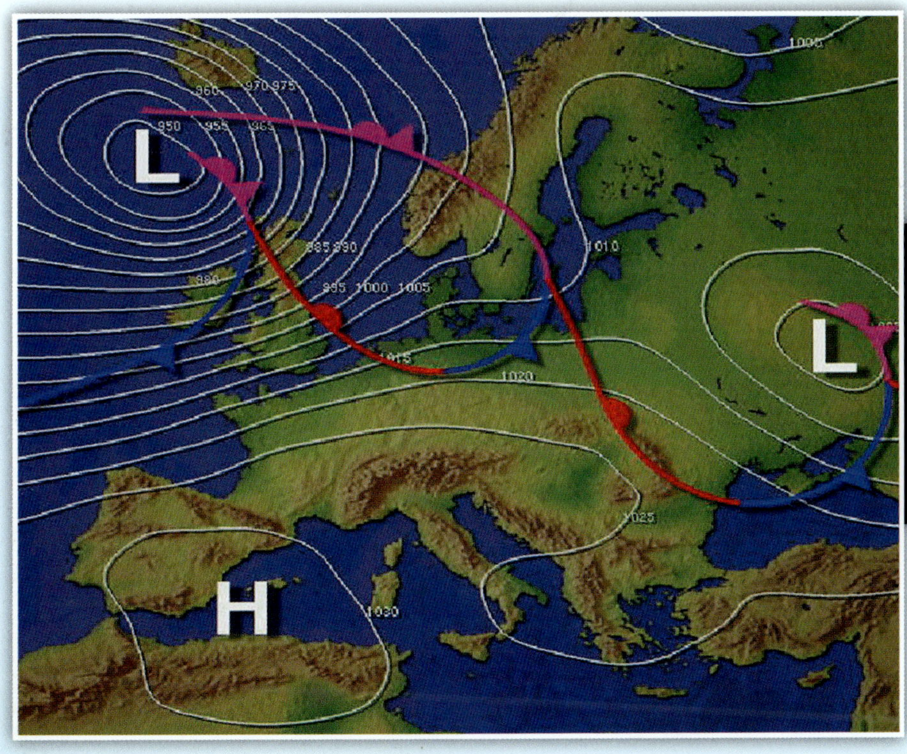

Reading a weather map can help you predict how the weather will change.

Weather tracking

For this activity you will need:

* A table that can be left outside
* A paint brush
* A wooden crate with slats to put over instruments
* A thermometer to measure temperature
* An **anemometer** to measure wind speed (see page 24)
* A wind vane to measure wind direction
* A **barometer** to measure pressure (see page 36)
* A rain gauge (see page 30)
* White paint
* A notepad and a pencil.

Steps to follow

1. Paint the outside of the crate white. A white box will reflect sunlight, so the inside does not heat up. This keeps extra heat away from your instruments, so your readings will be accurate.

2. Place your instruments (except for the **rain gauge**) on the table and place the crate upside down over your instruments. The slats will allow air to flow in, while the crate keeps the instruments protected from wind and sunlight.

3. Set up the rain gauge near the table.

4. Make daily measurements of the weather, one or more times a day, at the same time each day. Record the following: temperature, wind direction, wind speed, air pressure level, rainfall, current conditions (rainy, sunny, cloudy, etc.).

5. Compare your measurements with those reported in a daily newspaper, or on an online weather site. Write any differences between your measurements and the measurements you read in a notebook.

Date	Temperature	Wind direction	Wind speed	Air pressure	Rainfall	Current conditions

6

After one or two weeks, look for patterns in your records. What weather follows when air pressure is low or high? What happens when the wind changes direction? How do clouds change the temperature during the day? What happens when skies are clear at night?

7

Try to predict the day's weather based on your measurements. Make a note of your predictions, and the actual weather that you observed.

If you set up your own weather station, you can track changes in the weather.

What Will the Weather Be?

We watch the weather because it is important to our lives. **Satellites**, computers, and other instruments track changes in the **atmosphere** and predict the weather. We often know when a major storm approaches. Lives are saved because people are warned about tornadoes, hurricanes, and other powerful storms.

But will our weather change? Will there be more floods, storms, and droughts in the future?

Scientists think that powerful storms may become more common as temperatures rise.

Climate change

Many scientists are concerned that **climate** change will affect our weather. In the past, the climate of different regions of the Earth has changed from warm to cold and cold to warm. Places that were once dry became wetter, and wet places became drier.

During past ice ages, parts of North America, Europe, and Asia were covered with huge sheets of ice. There have been at least four ice ages in the past, the most recent one ending about 11,000 years ago. Around 65 million years ago, a changing climate may have led to the disappearance of the dinosaurs. An **asteroid** might have collided with the Earth, changing the temperature. The Earth has seen many natural changes in climate.

If you can predict the weather, you will know when you can enjoy the outdoors!

Global warming and weather

In the last few decades, people have caused climates to change at a fast pace. We use fuels such as gasoline and coal in our cars, factories, and to heat buildings. Burning fuels release carbon dioxide and other **gases** into the atmosphere. When people burn and clear forests to make large farms, the atmosphere also changes. Gases such as carbon dioxide trap heat. Because there are more of these gases now in the atmosphere, more heat is trapped close to Earth. The temperature of the Earth rises, causing global warming.

A warm atmosphere may cause more severe storms. **Ice caps** might melt and the oceans warm, causing flooding in coastal areas. Other places may have droughts. People are not certain how our weather will change as a result of global warming, but scientists think that powerful storms might become more common.

Glossary

air mass large body of air with the same temperature, pressure, and moisture content throughout

air pressure push made by the air in the atmosphere, on a particular area

anemometer instrument for measuring wind speed

asteroid irregularly shaped rock that orbits the sun

atmosphere mixture of gases that surrounds the Earth; the atmosphere has five layers

axis imaginary straight line from the North Pole to the South Pole, around which the Earth rotates

barometer instrument for measuring air pressure

cirrus cloud wispy white cloud that is high in the sky. Cirrus clouds are often made of ice crystals.

climate typical weather in a certain location over a long period of time

cold front boundary where a cold air mass pushes a warm air mass and replaces it

condensation process of changing from a gas to a liquid

cumulus cloud heaped, puffy cloud that looks like cotton. Cumulus clouds often mean fair weather, but can grow into rain clouds.

current movement of water or air in a particular direction

density mass per unit volume of a substance

doppler radar instrument for measuring speed

energy capacity to do work, for example, to provide light or heat

equator imaginary line around the Earth that is halfway between the North and South Poles. The sun's focus is greatest at the equator.

evaporation process of changing from a liquid to a gas without boiling

exosphere layer of the Earth's atmosphere, above the thermosphere and reaching into space

gas one of the three states of matter, along with liquid and solid

hemisphere half of the planet

high pressure high force made by sinking cool air

humid containing a lot of water vapor

ice cap covering of ice over a large area

infrared waves wavelengths that are given off by heated objects

isobar line on a weather map connecting points with the same air pressure

liquid one of the three states of matter, along with gas and solid

low pressure low force made by rising warm air

mesosphere layer of the Earth's atmosphere above the stratosphere

meteorologist person who studies processes in the Earth's atmosphere that cause weather conditions

moderate keep at a sensible limit. The ocean moderates the air so it does not get too warm or cold.

molecule smallest particle that a substance can be broken into while still keeping the properties of the substance

occluded front mass of warm air pushed off the ground when a cold front catches up to a warm front

ozone form of oxygen gas. The ozone layer in the atmosphere blocks harmful rays of the sun and is part of the stratosphere, above the troposphere.

photosphere intensely bright outer surface of the sun

precipitation any form of water that falls to the Earth, including rain, snow, and hail

rain gauge instrument used to measure rainfall

satellite machine that orbits around the Earth. Weather satellites take pictures and measurements that help meteorologists forecast the weather.

saturated air air that contains as much vapor as it can at a given temperature and pressure

season period of the year characterized by a certain pattern of weather. Seasons are caused by the tilt of the Earth as it revolves around the sun.

stratosphere layer of the Earth's atmosphere above the troposphere

stratus cloud flat, low clouds that form a blanket in the sky. Stratus clouds make the sky look gray, but do not usually bring rain.

thermosphere layer of the Earth's atmosphere above the mesosphere

troposphere layer of the atmosphere closest to the Earth. Weather occurs in the troposphere.

warm front boundary where a warm air mass pushes over a cold air mass and replaces it

water cycle the continuous movement of water from the Earth to the atmosphere and back to the Earth

water vapor gas form of liquid water

Find Out More

Books

Challoner, Jack. *Hurricane and Tornado: DK Eyewitness Books*. (New York: DK Publishing, 2004)

Cherry, Lynne. *How We Know What We Know About Our Changing Climate: Scientists and Kids Explore Global Warming*. (Nevada City, Calif: Dawn Publications, 2008)

Cosgrove, Brian. *Weather: DK Eyewitness Books*. (New York: DK Publishing, 2007)

Haslam, Andrew and Barbara Taylor. *Make it Work! Weather* (Chicago: World Book, 2004)

Woodward, John. *Nature Activities: Weather Watcher*. (New York: DK Publishing, 2006)

Websites

Edheads: Activate Your Mind!

www.edheads.org/activities/weather

Take an underground trip and learn to report and predict the weather.

NOAA Education: Weather

www.education.noaa.gov/sweather.html

Explore your planet through fun activities!

National Weather Service

www.nws.noaa.gov/outlook_tab.php

Read weather maps to find out about weather across the United States.

Southwest Florida Management District: Kids

www.swfwmd.state.fl.us/education/kids

Learn about the **water cycle**, hidden water, lightning, and more.

Websites continued

Space Weather Research Explorer

www.exploratorium.edu/spaceweather/index.html

Learn how weather in space can affect Earth.

Web Weather for Kids

http://eo.ucar.edu/webweather

Learn what makes weather wet and wild, and how to forecast the weather.

Weather Wiz Kids

www.weatherwizkids.com

Learn about hurricanes, tornadoes, clouds, winter storms, and more from a **meteorologist**.

Places to visit

Echo Lake Aquarium and Science Center
Leahy Center for Lake Champlain
1 College Street
Burlington
Vermont 05401
Tel: (802) 864-1848

www.echovermont.org

Be a guest meteorologist and report today's weather from the center's weather exhibit.

Rochester Museum and Science Center
657 East Avenue
Rochester, NY 14607
Tel: (585) 271-4320

www.rmsc.org

Lightning flashes and thunder crashes as you explore weather. Use computers to forecast the weather, and then record your forecast on camera.

The John C. Freeman Weather Museum
Weather Research Center
5104 Caroline
Houston, Texas 77004
Tel: (713) 529-3076

Become a weather wizard and learn about weather and **climates** around the world.

Index